Sir Francis Drake

Description of his landing at
Drake's Bay, Marin County,
California

June 17, 1579

Being an exact copy of parts of the
original report of this VOYAGE, in
his caravel the "GOLDEN HIND,"
including a description of the first
religious service in English ever
held in AMERICA, and also
the date of his departure for
ENGLAND, JULY 25th, 1579.

THE NARRATIVE

"In 38 Deg. 30 min. we fell with a conuenient and fit harborough and June 17 came to anchor therein. * * *

The next day, after our coming to anchor in the aforesaid harbour, the people of the countrey shewed themselues, sending off a man with great expedition to vs in a canow. * * *

The 3 day following, uiz., the 21, our ship hauing receiued a leake at sea, was brought to anchor neerer the shoare, that, her goods being landed, she might be repaired; but for that we were to preuent any danger that might chance against our safety, our generall first of all landed his men, with all necessary prouision, to build tents and make a fort for the defense of o u r selues and goods; and that wee might vnder the shelter of it with more safety (what euer should befall) end our businesse; which when the people of the countrey perceiued vs do-ing, they came down to vs, and yet with no hostile meaning or intent to hurt vs.   *   *   *

Their men for the most part goe naked; the women take a kinde of bulrushes, and kembing it after the manner of hemp, make them-selues thereof a loose garment, which being knitte about t h e i r middles, hanges downe a b out their hippes, a n d so affordes to them a couering.   They are very obedient to their husbands.

and e x c e e d i n g ready in all
seruices. * * *

Against the end of three daies
more (the newes hauing the while
spread itselfe farther, and as it
seemed a great way vp into the
countrie), w e r e assembled the
greatest number of people which
wee could reasonably imagine to
dwell within any conuenient dis-
tance round about. Amongst the
rest the king himselfe, a man of
goodly stature and comely person-
age, attended with his guard of
about 100 tall and warlike men,
this day, uiz., J u n e 26, came
downe to see vs. * * *

In the meane time the women
remaining on the hill, tormented
themselues lamentably, tearing

their flesh from their cheekes, whereby we perceiued that they were about a sacrifice. In the meane time our generall, with his companie, WENT TO PRAYER, AND TO READING OF THE SCRIPTURES, AT WHICH EX- ERCISE THEY WERE ATTEN- TIUE, and seemed greatly to be affected with it. * * *

This one thing was obserued to bee generall amongst them all, that euery one had his face painted, some with white, some blacke, and some with other col- ours, euery man also bringing in his hand one thing or other for a gift or present. * * *

Few were the dayes, wherein they were absent from vs, during and kingdome, both by the king and people, into h e r maiesties hands; together with h e r highnesse picture a n d armes, in a piece of sixpense currant English monie, showing itselfe by a hole made of purpose through t h e plate; vnderneath was likewise engrauen the name of our generall, etc. * * *

And now, as the time of our departure was perceiued by them to draw nigh, so did the sorrowes and miseries of this people seeme to themselues t o increase vpon them.

The 23 of July they tooke a sorrowfull farewell of vs, but being loath t o leaue vs, they presently ranne to the top of the hils

to keepe vs in their sight as long as they could, making fires before and behind, and on each side of them, burning therein (as is to be supposed) sacrifices at our departure. * * *

Not farre without this harborough did lye certaine ilands (we called them the ilands of Saint James), hauing on them plentifull a n d great store of seales a n d birds, with one of which wee fell July 24, whereon we found such prouision as might competently serue our turne for a while. We departed againe the day next following, uiz., July 25, 1579. * *

THEN AND NOW

"The 23 of July they toake a sorrowful farewell of us, making fires before and behind and on each side of them (as is supposed) sacrifices at our departure."

So wrote the ancient chronicler of the Drake expedition and sank to slumber while the Golden Hind plowed its way homeward. In a second (for in the great calendar) three centuries are but as seconds) those people who tormented themselves lamentably have vanished and the goodly country which Drake foresaw has come into its own. Home estates, farms and bungalows crowd its hills and valleys, and the spot has become a principality, tributary to a metropolis whose spires have risen like magic on what was the desolate sand spit to the south. The ancient chronicler was a worthy prophet. "A goodly country." No more beautiful spot than Marin County can be found in all the world lying so close to a metropolis. Neither London, Paris, Berlin, New York nor any other commercial center of which we know, has within one hour of the busy city streets such a variety of hills, mountains, valleys, cascades, streams and wooded trails as Marin County offers to San Francisco. In fact, here at one's very door is a mountain fastness possessing all of the beauties that an indulgent nature and climate has to offer.

MARIN COUNTY

Take boat from San Francisco and in half an hour one lands upon the selfsame peninsula where landed Drake. Explore inland as did he; but not as painfully as did he. Electric roads add to the pleasure and accessibility of this garden spot.

The boat itself lands one at Sausalito, a city of beautiful homes upon terraced hills and looking out across the great bay. Before it, in the waters of Tiburon, rests Belvedere, a dream island of homes.

Across the hills a vigorous walk lands one on Point Bonita where a picturesque lighthouse looks across the sea to China and Japan and sees close at hand that island which the brave voyagers of Drake called the "Iland of Saint James," "having on them plentiful and great store of seals and birds."

Again an electric train from Sausalito and one is at the foot of Mt. Tamalpais amidst the pines and redwoods. Trails wind through the woods up and up until one gains the crest of the divide. Far away one sees the silver of the Pacific; below, at one's very feet, the green of the Muir Woods where stand the forest monarchs which stood when Drake landed, and even in that day they were hoary veterans: along the crest lies West Point; and further still, at the summit of Mt. Tamalpais, the Tavern from which the whole

country lies spread out like a map.
One sees the gleam of San Francisco Bay, the dark green of the
forests, the lighter green of the
valleys and farms, the glitter of
the nestling towns, the slow crawl-
ing trains, a picturesque panorama
covering nine counties; and far
away, above the clouds Mt.
Diablo peering across the open to
its neighbor; the Sierra Nevadas
further to the east stately with its
crown of perpetual snow; and Mt.
Hamilton on whose summit the
Lick Observatory, that outpost of
Science watching with unerring
eye the procession of the Heaven-
ly Hosts.

Again the electric line to
San Rafael, the ideal city of
homes and rich of woods,

MARIN COUNTY

passing on the way station after station where happy throngs wait the returning city traveler and happy homes peer out from the green of the woods. Here no snow falls nor winter chills; nature is kindly and like a kindly mother gathers to herself t h e laughing children giving them their birthright of sunshine and fresh air. Who would shut within bricks and mortar these tender shoots when such a climate and country lies so close at hand?

Or, if one can spare the time, press even further. The railroad runs through farms and valleys, past bungalows a n d w o o d e d slopes beside the ever changing streams till it comes to rest at last into the heart of the woods. Take a motor stage. S wi n g through the hills where nature rests silent and undisturbed till at last you break forth upon a wide sweep of majestic ocean. It is the Pacific. Hour after hour you may skirt the cliffs; the restless ocean tosses far below you, its white fangs gleaming upon the dripping rocks while the stately forests stand upon the hills gazing calmly down upon this giant who would eat his way into their sanctuary. Along the rockbound shore busy fisheries thrive, their industry supplying a ready market close at hand. "A goodly country and fruitfull soyle, stored with many blessings fit for the use of man."

MARIN COUNTY

We have been hearing the roar of cannon and the call to arms as Christian man springs like a wier wolf at the throat of Christian man; from the city comes the cry of travail as the wheels of toil and care grind on and ever on. Are you weary of all this? Does it rest as a burden on your soul? Take boat with Drake. Plunge into this goodly country. The sweet air will bring you solace. The hills a n d forests will look upon you with so deep a calm that you will wonder at your restlessness. The seemingly great things of the city will shrink to littleness while mother nature rocks you on her bosom. The Land of Drake will welcome you. A n d when you leave, it will be you that will take sorrowful farewell, not those blessed inhabitants who s t a n d watching on the hills.

E. F. Green.

Ingram Content Group UK Ltd.
Milton Keynes UK
UKHW050919070723
424569UK00021B/96